THE RETURN

Charles Tomlinson

Oxford   New York

OXFORD UNIVERSITY PRESS

1987

Oxford University Press, Walton Street, Oxford OX2 6DP
Oxford   New York   Toronto
Delhi   Bombay   Calcutta   Madras   Karachi
Petaling Jaya   Singapore   Hong Kong   Tokyo
Nairobi   Dar es Salaam   Cape Town
Melbourne   Auckland
and associated companies in
Beirut   Berlin   Ibadan   Nicosia

Oxford is a trade mark of Oxford University Press

First issued as an Oxford University Press paperback 1987

British Library Cataloguing in Publication Data
Tomlinson, Charles
The return.—(Oxford poets).
I. Title
821'.914   PR6039.0349
ISBN 0-19-282079-6

Library of Congress Cataloging in Publication Data
Tomlinson, Charles, 1927–
The return.
I. Title.
PR6039.0349R4 1987   821'.914   87-7784
ISBN 0-19-282079-6

Set by Rowland Phototypesetting Ltd.
Printed in Great Britain by
J. W. Arrowsmith Ltd., Bristol

In memory of Jasper

ACKNOWLEDGEMENTS

ACKNOWLEDGEMENTS are due to the editors of the following: *Agenda, Between Comets, Cream City Review, The London Magazine, Mandeville's Travellers, Partisan Review, Pequod, Perfect Bound, Poetry Australia, Poetry (Chicago), Poetry Nation Review, Poetry Review, Sagetrieb, South West Review, The Times Literary Supplement, Vuelta* (Mexico), *Word and Image*. These poems first appeared in *The Hudson Review*: 'In the Borghese Gardens', 'Fountain', 'Between Serra and Rocchetta'. 'The Miracle of the Bottle and the Fishes' was commissioned by the Tate Gallery.

CONTENTS

In the Borghese Gardens   1
Revolution   2
Fountain   3
In San Clemente   4
In the Gesù   5
The Wind from Africa   6
The Return
    I The Road   7
   II Between Serra and Rocchetta   8
  III Graziella   10
   IV The Fireflies   11
Tyrrenhean   12
Travelling   13
Palermo   14
Catacomb   15
The Unpainted Mountain   17
The Miracle of the Bottle and
    the Fishes   18
Soutine at Céret   20
Self-Portrait   21
The Peak   22
Netherlands   23
Anniversary   24
In Memory of George Oppen   25
In Oklahoma   26
Interpretations   27
At Chimayó   28
The Tax Inspector   30
At Huexotla   31
Macchu Picchu   32
Winter Journey   33
Heron   36
The Harbinger   38
A Rose for Janet   39

February   40
Mid-March   41
From Porlock   42
The Well   43
Coombe   44
Hearing the Ways   45
What Next?   46
The Night Farm   47
The Hawthorn in Trent Vale   48
The Way Through   49
Night Ferry   50
Ararat   51

THE RETURN

IN THE BORGHESE GARDENS

for Attilio Bertolucci

Edging each other towards consummation
On the public grass and in the public eye,
Under the Borghese pines the lovers
Cannot tell what thunderheads mount the sky,
To mingle with the roar of afternoon
Rumours of the storm that must drench them soon.

Cars intersect the cardinal's great dream,
His parterres redesigned, gardens half-gone,
Yet Pluto's grasp still bruises Proserpine,
Apollo still hunts Daphne's flesh in stone,
Where the Borghese statuary and trees command
The ever-renewing city from their parkland.

The unbridled adolescences of gods
Had all of earth and air to cool their flights
And to rekindle. But where should lovers go
These torrid afternoons, these humid nights
While Daphne twists in leaves, Apollo burns
And Proserpine returns, returns, returns?

Rome is still Rome. Its ruins and its squares
Stand sluiced in wet and all its asphalt gleaming,
The street fronts caged behind the slant of rain-bars
Sun is already melting where they teem:
Spray-haloed traffic taints your laurel leaves,
City of restitutions, city of thieves.

Lovers, this giant hand, half-seen, sustains
By lifting up into its palm and plane
Our littleness: the shining causeway leads
Through arches, bridges, avenues and lanes
Of stone, that brought us first to this green place—
Expelled, we are the heirs of healing artifice.

Deserted now, and all that callow fire
Quenched in the downpour, here the parkland ways
Reach out into the density of dusk,
Between an Eden lost and promised paradise,
That overbrimming scent, rain-sharpened, fills,
Girdled within a rivercourse and seven hills.

I

REVOLUTION

PIAZZA DI SPAGNA

REVOLUTION it says
painted in purple along
the baluster of the Spanish
Steps and yes
you can return
by the other side
of this double stair
to where the word
is guiding you
(a little breathless)
down, up and back:
returning
you must run a ring
round the sun-browned
drop-outs
who litter the ascent:
their flights are inward
unlike these,
unfolding by degrees
what was once a hill,
each step a lip
of stone and what they say
to the sauntering eye
as clear as the day
they were made
to measure out and treasure
each rising inch
that nature had mislaid:
for only art
can return us to an Eden where
each plane and part
is bonded, fluid, fitting and
fits like this stair.

FOUNTAIN

Art grows from hurt, you say. And I must own
Adam in Eden would have need of none.
Yet why should it not flow as a Roman fountain,
A fortunate fall between the sun and stone?
All a fountain can simulate and spread—
Scattering a music of public places
Through murmurs, mirrors, secrecy and shade—
Makes reparation for what hurt gave rise
To a wish to speak beyond the wound's one mouth,
And draw to singleness the several voices
That double a strength, diversify a truth,
Letting a shawl of water drape, escape
The basin's brim reshaping itself to fill
A whole clear cistern with its circling calm,
And the intricacies of moss and marble
With echoes of distance, aqueduct and hill.

IN SAN CLEMENTE

What deer are these stand drinking at the spring?
Ask of the child the saint is carrying
Across a stream in spate. The steps that flow
Downwards through the sonorous dark beneath,
Should be a water-stair, for where they go,
A child that angels bring forth on the wall
Has lived a whole year on the ocean bed;
Then, down once more, and past the humid cave
Of Mithras' bull and shrine, until they lead
To a wall of tufa and—beyond—the roar
Of subterranean waters pouring by
All of the centuries it takes to climb
From Mithras to the myth-resisting play
Of one clear jet chiming against this bowl
In the fountained courtyard and the open day.

IN THE GESÙ

All frescoed paradise in adoration,
Saints choir the unanimity each atom feels,
And hearts that cannot rise to the occasion
Are spurned to earth beneath angelic heels.
This is the church triumphant, not so loving
As winged with a resistless certainty:
This is the despotism of the dove,
The empire of love without love's comity.

THE WIND FROM AFRICA

Wind ransacks every colonnade and corner,
Rattles the empty Coke cans, drags shutters back
To show St Peter's palled in dusty fog.
A cat creeps down the Campidoglio steps—
A stage-set fit to play the Roman on—,
Then blows to the stairfoot like a weightless rag
And slides and sidles out across the square.
Sahara's particles have greyed to London
The whole of Rome, complete with a paper snow-storm:
'And if it comes from Africa,' an old man says, 'why is the wind not
    warm?'

THE RETURN

to Paolo Bertolani

I THE ROAD

I could not draw a map of it, this road,
Nor say with certainty how many times
It doubles on itself before it climbs
Clear of the ascent. And yet I know
Each bend and vista and could not mistake
The recognitions, the recurrences
As they occur, nor where. So my forgetting
Brings back the track of what was always there
As new as a discovery. And now
The summit gives us all that lies below,
Shows us the islands slide into their places
Beyond the shore and, when the lights come on,
How all the other roads declare themselves
Garlanding their gradients to the sea,
How the road that brought us here has dropped away
A half-lost contour on a chart of lights
The waters ripple and spread across the bay.

Walking to La Rocchetta, thirty years
Would not be long enough to teach the mind
Flower by flower their names and their succession.
Walking to La Rocchetta, leave behind
The road, the fortress and the radar tower
And turn across the hill. From thirty years
I have brought back the image to the place.
The place has changed, the image still remains—
A spot that, niched above a half-seen bay,
Climbs up to catch the glitter from beyond
Of snow and marble off the Apennines.
But where are the walls, the wells, the living lines
That led the water down from plot to plot?
Hedges have reached the summits of the trees
Over the reeds and brambles no one cut.
When first I came, it was a time of storms:
Grey seas, uneasily marbling, scourged the cliff:
The waters had their way with skiff on skiff
And, beached, their sides were riven against stone,
Or, anchored, rode the onrush keels in air
Where hope and livelihood went down as one.
Two things we had in common, you and I
Besides our bitterness at want of use,
And these were poetry and poverty:
This was a place of poverty and splendour:
All unprepared, when clarity returned
I felt the sunlight prise me from myself
And from the youthful sickness I had learned
As shield from disappointments: cure came slow
And came, in part, from what I grew to know
Here on this coast among its reefs and islands.
I looked to them for courage across time,
Their substance shaped itself to mind and hand—
Severe the grace a place and people share
Along this slope where Serra took its stand:
For years I held those shapes in thought alone,
Certain you must have left long since, and then
Returning found that you had never gone.
What is a place? For you a single spot.
Walking to La Rocchetta we can trace
In all that meets the eye and all that does not
Half of its history, the other lies

8

In the rise, the run, the fall of voices:
Innumerable conversations chafe the air
At thresholds and in alleys, street and square
Of those who climbed this slope to work its soil,
And phrases marrying a tongue and time
Coil through the mind's ear, climbing now with us
Through orchids and the wild asparagus:
For place is always an embodiment
And incarnation beyond argument,
Centre and source where altars, once, would rise
To celebrate those lesser deities
We still believe in—angels beyond fable
Who still might visit the patriarch's tent and table
Both here and now, or rather let us say
They rustle through the pages you and I
Rooted in earth, have dedicated to them.
Under the vines the fireflies are returning:
Pasolini spoke of their extinction.
Our lookout lies above a poisoned sea:
Wrong, he was right, you tell us—I agree,
Of one thing the enigma is quite sure,
We have lived into a time we shall not cure.
But climbing to La Rocchetta, let there be
One sole regret to cross our path today,
That she, who tempered your beginning pen
Will never take this road with us again
Or hear, now, the full gamut of your mastery.

We cannot climb these slopes without our dead;
We need no fiction of a hillside ghosted,
A fade-out on the tremor of the sea.
The dead do not return, and nor shall we
To pry and prompt the living or rehearse
The luxuries of self-debating verse.
Their silence we inhabit now they've gone
And like a garment drawn the darkness on
Beyond all hurt. This quiet we must bear:
Put words into their mouths, you fail to hear
What once they said. I can recall the day
She imitated my clipped, foreign way
Of saying *Shakespeare*: English, long unheard,
Came flying back, some unfamiliar bird
Cutting a wing-gust through the weight of air
As she repeated it—*Shakespeare Shakespeare*—
Voice-prints of a season that belongs
To the cicadas and the heat, their song
Shrill, simmering and continuous.
Why does a mere word seem autonomous
We catch back from the grave? The wave it rides
Was spent long-since, dissolved within the tides
Of space and time. And yet the living tone
Shaped to that sound, and mocking at its own,
A voice at play, amused, embodied, clear,
Spryer than any ghost still haunts the ear.
The dog days, the cicada had returned
And through that body more than summer burned
A way and waste into its dark terrain,
Burned back and back till nothing should remain,
Yet could not dry the mind up at its source:
Clear as her voice-print, its unyielded force
Would not be shadowed out of clarity
Until the moment it had ceased to be.
Downhill, between the olives, more than eye
Must tell the foot what path it travels by;
The sea-lights' constellations sway beneath
And we are on the Easter side of death.

I have climbed blind the way down through the trees
(How faint the phosphorescence of the stones)
On nights when not a light showed on the bay
And nothing marked the line of sky and sea—
Only the beating of the heart defined
A space of being in the faceless dark,
The foot that found and won the path from blindness,
The hand, outstretched, that touched on branch and bark.
The soundless revolution of the stars
Brings back the fireflies and each constellation,
And we are here half-shielded from that height
Whose star-points feed the white lactation, far
Incandescence where the single star
Is lost to sight. This is a waiting time.
Those thirty, lived-out years were slow to rhyme
With consonances unforeseen, and, gone,
Were brief beneath the seasons and the sun.
We wait now on the absence of our dead,
Sharing the middle world of moving lights
Where fireflies taking torches to the rose
Hover at those clustered, half-lit porches,
Eyelid on closed eyelid in their glow
Flushed into flesh, then darkening as they go.
The adagio of lights is gathering
Across the sway and counter-lines as bay
And sky, contrary in motion, swerve
Against each other's patternings, while these
Tiny, travelling fires gainsay them both,
Trusting to neither empty space nor seas
The burden of their weightless circlings. We,
Knowing no more of death than other men
Who make the last submission and return,
Savour the good wine of a summer's night
Fronting the islands and the harbour bar,
Uncounted in the sum of our unknowings
How sweet the fireflies' span to those who live it,
Equal, in their arrivals and their goings,
With the order and the beauty of star on star.

TYRRENHEAN:
FROM THE TRAIN

Not recognizing the thing it was, I caught
For a moment an elsewhere in the view:
It seemed a luminous steppe, a plain of blue
Had risen up suddenly beneath the sky—
A grass ocean, yet neither grass nor wave
Stirred on this calm that would persuade the eye
It was as deep and changeless as the grave.
This Blue Grass was a country of the mind,
And yet its sheer impossibility
Brought home unmediated sea, a crop
That had sprung up overnight but always been
There rustling and ready to be seen,
Though the eye rode past it at a glance
Filled with the certainty that sea was sea.

TRAVELLING

POEM ENDING WITH A LINE OF EMERSON

The storm, as it closed in, blackening round the train,
   Flushed back the yellows of the passing field
To twilit vanishings, as if the sense
   Faltering, had gone out. Rain
Rushed the sleeping ear into wakefulness;
   At every window it crashed and rang
Like gravel against the glass, like clashed keys flung
   As, all forms gone, we swung through the daylight dark
And on, in a tumultuous privacy of storm.

PALERMO

The road to Palermo flows through a tunnel of trees—
    An asphalt waterway, a discovery
The car-lights keep extending
    Ahead of themselves. Fencing back
Mountains, plains and sea, this dense
    Illusion of a forest seems more real
Than the vista replacing it, a perspective
    Down decaying streets that stops
At the water's margin beneath the impartial
    Cranes conferring together above roof-tops.

CATACOMB

A capuchin—long acquaintance with the dead
    Has left him taciturn—stands guard
At gate and stairhead. Silent, he awaits
    The coin we drop into his dish, and then
Withdraws to contemplation—though his eye
    Glides with a marvellous economy sideways
Towards the stair, in silent intimation
    You may now descend. We do—and end up
In a corridor with no end in view: dead
    Line the perspective left and right
Costumed for resurrection. The guidebook had not lied
    Or tidied the sight away—and yet
Eight thousand said, unseen, could scarcely mean
    The silence throughout this city of the dead,
Street on street of it calling into question
    That solidity the embalmer would counterfeit.
Mob-cap, cape, lace, stole and cowl,
    Frocked children still at play
In the Elysian fields of yesterday
    Greet each morning with a morning face
Put on a century ago. Why are we here?—
    Following this procession, bier on bier
(The windowed dead, within), and those
    Upright and about to go, but caught
Forever in their parting pose, as though
    They might have died out walking. Some
Face us from the wall, like damaged portraits;
    Some, whose clothing has kept its gloss,
Glow down across the years at us
    Why are you here? And why, indeed,
For the sunlight through a lunette overhead
    Brightens along a sinuous bole of palm:
Leaves catch and flare it into staring green
    Where a twine of tendril sways inside
Between the bars. Light from that sky
    Comes burning off the bay
Vibrant with Africa; in public gardens
    Tenses against the butterflies' descent
The stamens of red hibiscus. Dead
    Dressed for the promenade they did not take,
Are leaning to that light: it is the sun
    Must judge them, for the sin

Of vanity sits lightly on them: it is the desire
    To feel its warmth against the skin
Has set them afoot once more in this parade
    Of epaulette, cockade and crinoline. We are here
Where no northern measure can undo
    So single-minded a lure—if once a year
The house of the dead stood open
    And these, dwelling beneath its roof,
Were shown the world's great wonders,
    They would marvel beyond every other thing
At the sun. Today, the dead
    Look out from their dark at us
And keep their counsel. The capuchin
    Has gone off guard, to be replaced
By a brother sentry whose mind is elsewhere—
    Averted from this populace whose conversion
Was nominal after all. His book
    Holds fast his eyes from us. His disregard
Abolishes us as we pass beyond the door.

Palermo

THE UNPAINTED MOUNTAIN

There was a storm of wind and light. The rock
Darkened to shadow and then flared to chalk.
All of the mountain seemed disparity
That I could neither think to one, nor see
Save as a tomb of stones, a livid chine
That blocked back distance. The horizon line
Straddled a mass of salt, of quarried snow.
Where were the folds, the facets that we knew—
The confirmation of the view we wanted,
Where was the mountain that the painter painted?
It was the light, you said, had changed it all,
Sapped out the warmth and left this interval
Of limestone where a pulse no longer beat
Between the scintillations of a heat
That wove the slope, the summit and the glance
Together in one dancing radiance.
Two tones—between-tones, both—pink-grey, green-black,
Flushed through the scene whenever sun came back:
The whole place hovered, image upon image:
A flank of marble or a crumpled page,
The bushes dotted a rucked leopard skin:
What we were seeing was unpainted mountain.
A blue recession, then, of mass gone-by,
Took a new colour out of space and sky:
The height was shepherding its shadows in
Out of the wind-torn vistas of the plain.
Distance and hills had hidden it from view
Till, unforeseen, a twist of road let through
An image that we knew from foot to crown,
As suddenly a seam of light ran down
The western slope and left it separate
Before that place where, soon, the sun would set
Though out of eye-shot, and already sent
A glow and earnest up the whole extent
Of Mont Ste-Victoire, altar more than hill.
Before us, now, new shapes began to fill
The poised horizon where the Luberon
Massed to our approach as night came on.

THE MIRACLE OF THE BOTTLE
AND THE FISHES

I

What is it Braque
would have us see in this
piled-up table-top of his?

One might even take it for
a cliff-side, sky-high
accumulation opening door on door

of space. We do not know
with precision or at a glance
which is space and which is substance,

nor should we yet: the eye must stitch
each half-seen, separate
identity together

in a mind delighted and disordered by
a freshness of the world's own weather.

II

To enter space anew:
to enter a new space
inch by inch and not
the perspective avenue
cutting a swathe through mastered distance
from a viewpoint that is single:
'If you painted nothing but profiles
you would grow to believe
men have only one eye.'
Touch must supply
space with its substance and become
a material of the exploration
as palpable as paint,
in a reciprocation where
things no longer stand
bounded by emptiness: 'I begin,'
he says, 'with the background
that supports the picture
like the foundation of a house.'

III

These layered darknesses
project no image of a mind
in collusion with its spectres:
in this debate
of shadow and illumination fate
does not hang heavily
over an uncertain year
(it is nineteen-twelve) for the eye
leaves fate undone
refusing to travel straitened
by either mood or taken measure:
it must stumble, it must touch
to guess how much of space
for all its wilderness
is both honeycomb and home.

SOUTINE AT CÉRET

The mistral is tearing at trunks: they flow in it.
    Cypresses lean in file, nodding and bowing
With the ceremonious automatism of trees.
    The heaped-up land hoists dwellings
Above rock-packs stacked
    And collapsing sideways and yet
Holding on. Here, olives angle themselves
    To cling to slopes, bending their backs
Like the old man climbing village steps:
    His legs strengthlessly weigh down
Drawn to the inertia of the stone in walls
    That circle and scale these hills. You feel
The mistral could billow out the houses
    With their windows and their doors
Into funfair mirror grimaces.
    The painter chose his landscape all too well
To fix the flux and turmoil of his hell,
    Deaf to the steady counsel of the rocks
And their refusal, anchorages firm,
    To liquefy to the impasto of his brushstrokes.

SELF-PORTRAIT

Grey on white, the pencil congregates
　　Its immutable wisps, its flecks of form:
He is drafting a portrait of himself
　　As someone else—sheer image
Without biography. Inhabiting the skin
　　No longer from inside, he declares it there
As pure stranger, a bush of lines
　　Growing before his eyes, until
There stares at him out of its own profusion
　　That other awakened from himself and slowly
Across the space consenting recognition.

THE PEAK

Descending, each time we came to where
    The snail-shell road-bend turned
Back circling on itself, we faced
    Into the peak once more. Ice
Had scooped and scraped the rock
    That climbed up to it, rivered
By waters no foot had walked beside.
    The mountain-head possessed a face
Of snow—blown into profile by the winds
    Travelling across it, blindly duning there
This human shape. The peak beyond
    Rose scoured by those same currents
And blown clear: its seamed rock
    Worn by no human wrinkles, stood
At a frontier. Rounding the mountain, we
    Stood with it: the descent, the repetition
As that farness fronted us yet again
    Might have been a dream or a damnation,
But the lowlands opened to receive us,
    Brought us the first sun free of mountain shadow
And the demarcation of ploughland, vineyard, meadow.
    From above the snow-line, and above the snow,
Something was tracking us, measuring our return
    Past the stone certitude of barn on barn.

NETHERLANDS

The train is taking us through a Mondrian—
　　The one he failed to paint. Cows
Keep moving along the lines
　　Of dyke and drain, the glinting parallels
And the right-angles of a land hand-made.
　　True: curvature is no feature of this view,
Yet why did the sky never cause him to digress
　　With its mile-high cloud mountains
Pillowed and piled over hill-lessness?
　　Flying between the two, go heron and gull
Hunters, haunters of every channel.
　　There are no verges unplanted, no acres spare:
Water continues accompanying our track,
　　Cows graze up to the factory windows and we are there.

ANNIVERSARY

for Beatrice

Over our roof the planes climb west.
   I caught today far out in space
A jet gone inching up the stratosphere
   Vertically from a stand of trees,
Over, then down to disappear
   Nose-diving the horizon. On high
An arc of vapour scored the shape
   One could take the planet's improbable measure by;
Then, as it began to fray and fade
   Bits of the bridge came floating down the sky
Combed by the wind. Yet out of view
   That arc went rounding on and on
And, half a world away, took also you
   Under its wing, reminding this spring day
What day it was by writing it overhead
   In a script that only April and you can read.

IN MEMORY OF GEORGE OPPEN

We were talking of O'Hara.
'Difficult', you said
'to imagine a good death—*he died*
quietly in bed, in place of:
he was run down
by a drunk.' And now, your own.
First, the long unskeining year by year
of memory and mind. You 'seemed
to be happy' is all I hear.
A lost self does not hide:
what seemed happy was not you
who died before you died. And yet
out of nonentity, where did the words
spring from when
towards the end you told
your sister, 'I don't know
if you have anything to say
but let's take out all the adjectives
and we'll find out'—the way,
lucidly unceremonious,
you spoke to her in life and us.

IN OKLAHOMA

Driving to Anadarko was like flight,
Gliding and grazing the surfaces of land
That flowed away from one secretively,
Yet seemed—all sparsely-treed immensity—
To have nothing to hide. Only the red
Declared itself among the leeched-out shades,
Rose into the buttes, seeped through the plain,
And left, in standing pools, one wine-dark stain.
The trees, with their survivors' look, the grasses
Yellowing into March refused their space
Those colours that would quicken to the ring
Of the horizon each declivity
And flood all in the sap and flare of spring.
The wideness waited. Sun kept clouded back.
An armadillo, crushed beside the road,
Dried out to a plaque of faded blood.
Here, fundamentalists have pitched their spires
Lower than that arbiter of wrath to come—
The tower of the tornado siren
Latticed in iron against a doubtful sky.

Anadarko is an Indian site. Near here the Tonkawa Hills Massacre took place in 1864.

26

INTERPRETATIONS

Distinctive, those
concretions known
in Oklahoma as 'rose
rocks'—an allusion
to their red-brown sand-
colour and
similarity to a rose
in full bloom. Petals?
Clusters of barite
crystals are what they are—
the rose-shape made
by the growth of barium
as a divergent cluster
of blade on blade.
The rosettes fed
on an ancient red
sandstone—the host
rock whose colour
they acquired as they
lost their own:
quartz sand-grains
bonded together to become
$BaSO_4$
and await the rigours and the rains
of two million and more
years to petrify then expose
the rose rock or barite rose
in positive relief.
To the Indian eye
those years brought forth
such blood-bonded
and bunched tears as show
a grief of dispossession
no rocks or rock rose
forming could foreknow.

AT CHIMAYÓ

The sanctuary was begun in the New Mexico of 1813 by one Don Bernardo Abeyta. It is a low-lying church of cracked adobe adjoining the chapel of the Santo Niño Perdido, the Lost Child. In capital letters, pinned to the door, hangs a warning:

NO FOOD
DRINK OR
PETS INSIDE
THE CHURCH

This, as we discover, is a place of notices, messages and names. Particularly names: the givers of ex-votos have inscribed theirs; those who believe it was this place brought about a cure for themselves or their relations, have written signed letters to say so and these are duly exhibited within; the saints are labelled for us—among them, the less familiar San Calletano and San Martín de Porres. Christ is not simply Christ, but Nuestro Señor de Esquipulas.

As we go in through the vestibule the custodian is saying to a travelling salesman that, yes, she will take two dozen. We do not know whether she is speaking of the ex-voto images that are for sale, or the layettes for expectant mothers, blessed on behalf of the Santo Niño. Not only is this place associated with the Santo Niño but with healing, and beside the altar of the Niño there is a hole in the floor from which the miraculous earth is scooped up as a medicament.

We push open the door into the church. It bears another notice, complete with Spanish accent and intonation:

DONT LIVE THIS DOOR. OPEN.
PLEASE

A cloying smell of melting wax from the candles inside. The great vigas overhead and the rough, sturdy walls compact the silence. There is a gaiety about the images: a sculpted Christ wears bright mocassins, a painted St Michael dances on the dragon he has overcome.

The chapel of the Niño also has its notice:

HELPUS
KEEP A
CLEAN SCENE

—for many pilgrims come here and perhaps the excluded food, drink and pets somehow find their way inside. After all, what could be more rational, if one has a sick pet, than to bring it in to the source of healing? There are pilgrims here now. They file intently past the bright-red carving of San Rafael holding his fish, into the pósito—the room where the healing earth lies, and where a crucified Christ hangs from his cross wearing a baby-dress of turquoise-coloured rayon decorated with nylon lace and plastic roses. Innocence of taste possesses its own fecundity to which these crowded walls bear witness, covered, as they are, with letters of thanks, photographs, crutches, Leonardo's Last Supper in several reproductions, a Raphael Madonna, pictures of Christ from paint-by-numbers sets, an airforce uniform (a sergeant's), keys, a penned description of the sanctuary, beginning, 'In a valley protected by wild berrytrees', a poem—'The twinkle of a stary stary night'—by G. Mendosa of Las Cruces, a framed portrait of the black-faced Guadalupe Virgin which also contains those of the donors in passport-size snapshots. In the midst of all this sits the effigy of the Santo Niño de Atocha in his wooden stall, cloaked in green and white, his cherubic but sallow face shadowed by a cockaded hat of seventeenth-century cut. There is something dandyish about his attire, but the rope of turquoise beads he wears round his neck has been broken. Someone has cobbled it together with a plastic-coated hairpin. It serves to display a tiny white cross with a minute Christ on it.

In this fecund chaos the messages on the walls carry a sole discordant note, a confession of waste. It is written by a prisoner, still in gaol, who admonishes himself with a cross composed out of two words:

G
M O M
D

The others leave crutches, thanks for cures or the pictures they have painted. He is the one sinner to confess his faults, and the thought of them jars his prose into unpredictable clusters of rhyme: 'I've wasted my life and its cost me my family and friends . . . No longer with a home not even a place to roam and this cell has become my domain. I know that its blame for bringing shame to my name and now I must part with my time.' 'Bringing shame to my name . . .': among all these Mendozas, Gonzalezes, Medinas, Antonios, Serafíns, Geias, he alone guards his anonymity. In this place of names, he is the only one to realize that to use his name would be a sort of blasphemy and that here he must forfeit it.

Driving back through the dusk I find it is his unsigned letter keeps returning to mind, outdoing the presence of those garish saints. And I wonder from what source a feudal word like 'domain' came to him in his cell where, King of Lackland, he is monarch of all he surveys. Perhaps, turning it over on his tongue, he tastes anew each time the lost liberty of these vistas, this unfencable kingdom of desert and mountains.

THE TAX INSPECTOR
at Tlacolula

I had been here before.
I came back
to see the chapel
of hacked saints.
It was shut.
A funeral filled
the body of the church:
small women with vast lilies
heard out the mass: the priest
completing communion
wiped wine
from his lips and from
the gold chalice
which having dried
he disposed of: the event
was closed. The organ
whose punctuations
had accompanied the rite
broke into a waltz
and the women
rose and the *compañeros*
de trabajo of the dead man
shouldered the coffin forth
to daylight. The waltz
seemed right as did
the deathmarch, the woe
of the inconsolable brass
preceding to the *campo santo*
the corpse, the women
and the *compañeros*
who sweated from street to street
under the bier,
swaying it like a boat.
And this was the way—
a banner declaring
what work he and his *compañeros*
had once shared—
the tax inspector,
ferried across on human flesh,
was borne to burial.

AT HUEXOTLA

Tall on its mound, el Paupérimo—
the poorest
church in Mexico
and the smallest.

It was not the sight
but the sound of the place
caused us to quicken our step
across the intervening space

between us and it—
such skeins, scales, swells
came from each bell-tower
though not from bells.

Who would compose
a quartet for flutes?—and yet
that was the music
rose to assail us.

A minute interior:
sun on the gold:
flute-timbre on flute
still unfolded there.

Flanking the altar,
caged birds hung,
the alchemy of light transmuting
gold to song:

for it was the light's
reflection had set
those cages in loud accord
and only night would staunch it.

MACCHU PICCHU

All day, the weight of heat and then
    Evening brings in the thunder-heads:
A moving mountain leads their cavalcade
    Of silent herds, decaying and re-forming,
And the mountain, too, toils, trails
    Across the view of empty upper-sky
A whole high geography: foot-hills
    Hollows, vales and forests follow
The world-in-making of this awakened height
    That seeps up massively and darkly clear,
Through the more—or is it less—than human light,
    Like an inkblot spread-out magnified:
Forest climbs with the piling crag:
    The single bird that dips before it
Seems astray from there and flies to say
    That Macchu Picchu has been dispossessed
Even of its houses, its stone shells'
    Pure prospect of a dwelling place
And the storm that is rising will efface the rest.

WINTER JOURNEY

I

When you wrote to tell of your arrival,
    It was midnight, you said, and knew
In wishing me *Goodnight* that I
    Would have been long abed. And that was true.
I was dreaming your way for you, my dear,
    Freed of the mist that followed the snow here,
And yet it followed you (within my dream, at least)
    Nor could I close my dreaming eye
To the thought of further snow
    Widening the landscape as it sought
The planes and ledges of your moorland drive.
    I saw a scene climb up around you
That whiteness had marked out and multiplied
    With a thousand touches beyond the green
And calculable expectations summer in such a place
    Might breed in one. My eye took in
Close-to, among the vastnesses you passed unharmed,
    The shapes the frozen haze hung on the furze
Like scattered necklaces the frost had caught
    Half-unthreaded in their fall. It must have been
The firm prints of your midnight pen
    Over my fantasia of snow, told you were safe,
Turning the threats from near and far
    To images of beauty we might share
As we shared my dream that now
    Flowed to the guiding motion of your hand,
As though through the silence of propitious dark
    It had reached out to touch me across sleeping England.

Alone in the house, I thought back to our flood
    That left not an inch of it unbaptised
With muddy flux. Fed by the snow-melt,
    The stream goes lapping past its stone flank now,
And the sound beneath all these appearances
    Is of water, close to the source and gathering speed,
Netting the air in notes, letting space show through
    As sound-motes cluster and then clear
Down all its course, renewing and re-rhymed
    Further-off from the ear: I listen
And hear out what they have to say
    Of consequence and distance. That night
The wave rose, broke, reminded us
    We cannot choose the shape of things
And must, at the last, lose in this play
    Of passing lights, of fear and trust:
Waiting, as I wait now, I wish you could hear
    The truce that distils note-perfect out of dusk.

III

I must tell you of the moon tonight
    How sharp it shone. You have been gone three days.
Cold burnt back the mist. The planetarium
    Set out in clarity the lesson of the sky—
Half lost to me: I thought how few
    Names of the revolving multitude I knew
As they stood forth to be recognized. I saw
    The plough and bear—were those the Pleiades?
A little certainty and much surmise, what is the worth
    Of such half-recognitions? They must be
For all their revelation of one's ignorance
    Worth something—let me say this at least:
Though bidden by darkness to the feast of light,
    I came as one prepared, and what I could not name
Opening out the immensity flame by flame
    Found me a celebrant in the mass of night,
Where all that one could know or signify
    Seemed poor beside the reaches of those fires,
The moon's high altar glittering up from earth,
    Burning and burgeoning against your return.

I lay the table where, tonight, we eat.
 The sun as it comes indoors out of space
Has left a rainbow irising each glass—
 A refraction, caught then multiplied
From the crystal tied within our window,
 Threaded up to transmit the play
And variety day deals us. By night
 The facets take our flames into their jewel
That, constant in itself, burns fuelled by change
 And now that the twilight has begun
Lets through one slivered shaft of reddening sun.
 I uncork the wine. I pile the hearth
With the green quick-burning wood that feeds
 Our winter fires, and kindle it
To quicken your return when dwindling day
 Must yield to the lights that beam you in
And the circle hurry to complete itself where you began,
 The smell of the distance entering with the air,
Your cold cheek warming to the firelight here.

HERON

Metamorphosed
by a god
you could not flow
hunchback into ballerina
as swiftly as the stiff-
legged heron, frozen now
—a pond-side
garden bronze
the flash of fish
will bring to life once more.

That kink
in the neck
all but disappears as he
stretches it to drink
through the needle-fine
now slightly open
beak: they tell
how a heron hunted
can transfix
the falcon with that bill.

An aristocrat?
The moor-hen
waddles demotic
by this mincing stalker who
is weightless like his shadow;
however, you
must not judge
his politics or
his ancestors
by the way he walks.

For who would guess
that he, content
with solitariness
will nest
in the trees alongside
other species,
or that one so studious
of quiet
united with his kind
turns all at once gregarious?

THE HARBINGER

A peregrine peers in
at a November window:
in air he could see his prey
a hundred feet below

but misses me
in this glazed interior,
looks through it and beyond
to the darkness where

—a sight past seeing—
he might in one
undistracted stare
gaze prey into being:

then he is gone
this sharp-faced harbinger:
there follow
flurries of a fine snow—

an infinitesimal
and delicate feathering,
mazing the space he now
harries his cold-dazed quarry in.

A ROSE FOR JANET

I know
this rose is only
an ink-and-paper rose
but see how it grows and goes
on growing
beneath your eyes:
a rose in flower
has had (almost) its vegetable hour
whilst my
rose of spaces and typography
can reappear at will
(your will)
whenever you repeat
this ceremony of the eye
from the beginning
and thus
learn how
to resurrect a rose
that's instantaneous
perennial
and perfect now

FEBRUARY

In the month-long frost, the waters
    Combing the detritus that clogs a stream,
Leave gleaming in their wake these twists of glass,
    Caught crystals, petal and frozen frond:
At night, if you could fly and sweep
    With the owl's deep stare the valley reaches,
You would see in each water way
    These barbed garlands glistening back
The light of an oval moon, whose full is failing,
    That must pull awry the brilliant symmetry of day
On freezing day, and gather up at last
    These gauds that distract the owl's encompassing eye.

MID-MARCH

I hung the saw by its haft from the saw-horse end
Pointing away from me into the wind
So that the blade should injure no one: the current
Running against the saw-teeth headlong, sent
A thin cry up, half-way to music, a high
Metallic song and severing of air from air,
A prelude to the night's frost, as it were.
The in-between season vibrated from those notes.
For weeks of immoveable snow only a gong could float
The appropriate brooding major across the land
To summon the storms in. This wirey band
Of reverberations vanished skywards, or perhaps
The March moon's crescent of peel is a snapped-
Off fragment of one of them solidified
On the freezing blue of a darkened sky.

FROM PORLOCK

Winding on up by the public way
  The roar of descending water in his ears
From the torrent that runs counter to his climb,
  This person does not pause to investigate
The sheen, the shimmer at the edge
  Of visibility, or the sway and glint
Off the new mintings of metallic sea
  Down back below him. His eye
Is elsewhere than the spindly trees
  Wooding the gullies, writhen by their growth
Into such shapes as (judging by that look)
  Might figure forth his mind into a book,
Its script all knots and tilting stems
  Huddled within sheer margins. That wood
Of his can never emerge as trees
  But logged and ledgered. His mind
Is on business that it leans towards
  Though leans less now that the summit
And the moor itself have both been breasted
  And the soundshape of the skylarks' kingdom
Is ringing above him—a dome
  It would daze the eye to climb up into
But not the ear. His ear is listening
  Through and through the house whose door
He hammers at. His knock gone probing into every room,
  He has the dream by the root, he has it out,
Has now what, unknowing, he came for,
  And the larks suspended in their dome behind him,
Hears the steps of the dreamer approaching across the floor.

THE WELL

for Norman Nicholson

We loosen the coping-stone that has sealed for years
    The mid-field well. We slide-off this roof
That has taken root. We cannot tell
    How the single, pale tendril of ivy
Has trailed inside nor where
    In the dark it ends. Past the dark
A small, clear mirror sends
    Our images back to us, the trees
Framing the roundel that we make,
    A circling frieze that answers to the form
Of this tunnel, coiled cool in brick.
    We let down the plumbline we have improvised
Out of twine and a stone, and as it arrives,
    Sounding and sounding past round on round of wall,
Our images liquidly multiply, flow out
    And past all bounds to drown in the dazzle
As a laugh of light runs echoing up from below.

COOMBE

The secrecy of this coombe is weighted through
    With the pressures of the land that does not show
Over its ridge—the massing of the moors,
    The withstanding cliff and the inland sweep
And drop whose encompassing granite hand
    Extends us the deep lines of its palm
Through softer soils that a river
    Silvers and darkens between. Climb
To the crest and the river has lost itself
    Down in the leafed-round dip and now
Dartmoor is shouldering up against the sky
    Its stone-age pastures and the silences
It kept from the Romans with. In a buzzard's eye
    It might all lie one map, but we
Take in our territory by inches then by bursts,
    More like that heron who stands, advances, stands
Firm in the sliding Torridge that divides
    The sheer of the woodslope from the packed cornland.

HEARING THE WAYS

Stream beds are pouring
a week's rain through—
hill-race into hollow,
mill-race out of view.

Under a cleared sky
you can hear the ways
the waters are steering
and measure out by

the changes of tone
their purchase on place,
catch the live note off stone
in the plunge of arrival.

The closed eye can explore
the shapes of the vale
as sure as the Braille
beneath a blind finger.

In all its roused cells
the whole mind unlocks
whenever eye listens,
wherever ear looks.

WHAT NEXT?

As the week-long rain
washed at the slope
the land began
to slip jerkily
down—not all at once
but like the several
frames of an unsteady
film that then
came right
unexpectedly, to go on
pouring forth its smooth
successive images
of rain, mud, rain
until the film again
(a montage unintended)
cut and the whole
slope was coming away
and the image
of a tidal wave of clay
in which there hung
a crumbling island with the trees
still upright on it
buried the road and left us
under our oilskins
high and dry
to retreat into
a perplexed nervous domesticity
and ask ourselves
as we regained our hill-top house
what next . . . ?

THE NIGHT FARM

It seemed like a city hidden in the hill,
And this the first house with its flaring panes—
A forge, it might be, from which the fire pulsed out
Above the steep descent of streets whose veins
Of light wound down into the hill heart.
But the beams that had brought this hidden town
To birth in thought were also telling
Of the ardent geometry of dwelling
And the purity of the dark from which they shone.

THE HAWTHORN IN TRENT VALE

After fifty years, the hawthorn hedge
    That ran through the new estate
Still divides the garden ends, resists
    With wounds to wrist and elbow every move
To fell it or constrain. This ghost
    From a farm now gone, remains to haunt
And prick the sleep of gardeners dreaming ill
    Of the one unaccommodating dendrophil
Who has tenderly let his portion swell into tree entire,
    Green fire and blossom fed
From the darkness under bed and masonry.

THE WAY THROUGH

How deep is the sea-cave runs beneath the cliff?
    Less deep than the reflection on its floor:
For the sky comes into the cave to occupy
    A rock pool—sea that does not reflect the sea—
And this image of the whiteness of the sky
    Tinged by the pale green of the roof, the walls,
Reads like a gap in the floor of rock
    That you could fall through. The image refuses to believe
You are more solid than this downward way,
    This crevice that speaks of falling day by day by day
Through a space which does not yet exist,
    Abiding the note that will shatter glass and rock
As Dies Irae widens the last crevasse.

NIGHT FERRY

Where does the ripple in the sky begin?
Behind the mountains holding the waters in.
You'd think the ripple on the water spread
Through rock and pine, vibrating overhead
In one continuous circling-out of power,
From whitening wake through darkness to the stir
Of cloud that, wave on wave, drinks up the last
Suffusions of the sunlight without haste.
Then night is on it all. The parallels—
Two fraying lines the eyes can scarcely tell—
Narrow in foam towards a far pulsation
That shapes the wharf, the tiny constellation
Our stern is thrusting from us. Now we go
Darkness on either side and dark below,
Into the grasp of the expanse and deep;
Yet summit snows still glow beyond the sleep
Of other tints—the strait and shore, one shade—
Holding the space above us open, blade
On jagged blade, and cut against the sky
A frontier for us. Dark must liquefy
Even that height and as the peaks go down
The last reflected light fade, founder, drown.
Air seems coldest then, catching the breath away
In one extinction of the blood and day
As the north brings down the ether from each peak unseen,
Driving the watcher on the deck within.
We crouch in the body's half-way house, and yet
Our travelling lights caught up into the net
The waters cast around us, swaying shoals
Bulge at the interlacings, dolphin schools
Of light break from the water-mesh they make
And ride up along the sides from prow to wake.
Cold holds the boat. One thin wall keeps us dry.
Little unseams dark pines from dark of sky
And we hang in the balance of fathoms, chart and stars
Where mountains on mountains stand round us and only the
    water stirs.

ARARAT

We shall sleep-out together through the dark
The earth's slow voyage across centuries
Towards whatever Ararat its ark
Is steering for. Our atoms then will feel
The jarring and arrival of that keel
In timelessness, and rise through galaxies,
Motes starred by the first and final light to show
Whether those shores are habitable or no.

OXFORD POETS

Fleur Adcock
Yehuda Amichai
James Berry
Edward Kamau Brathwaite
Joseph Brodsky
Basil Bunting
D. J. Enright
Roy Fisher
David Gascoyne
David Harsent
Anthony Hecht
Zbigniew Herbert
Thomas Kinsella
Brad Leithauser
Herbert Lomas
Medbh McGuckian
Derek Mahon

James Merrill
John Montague
Peter Porter
Craig Raine
Tom Rawling
Christopher Reid
Stephen Romer
Peter Scupham
Penelope Shuttle
Louis Simpson
Anne Stevenson
Anthony Thwaite
Charles Tomlinson
Andrei Voznesensky
Chris Wallace-Crabbe
Hugo Williams